Logan,
Enjoy climbing trees
and have a happy
year.

[signature]

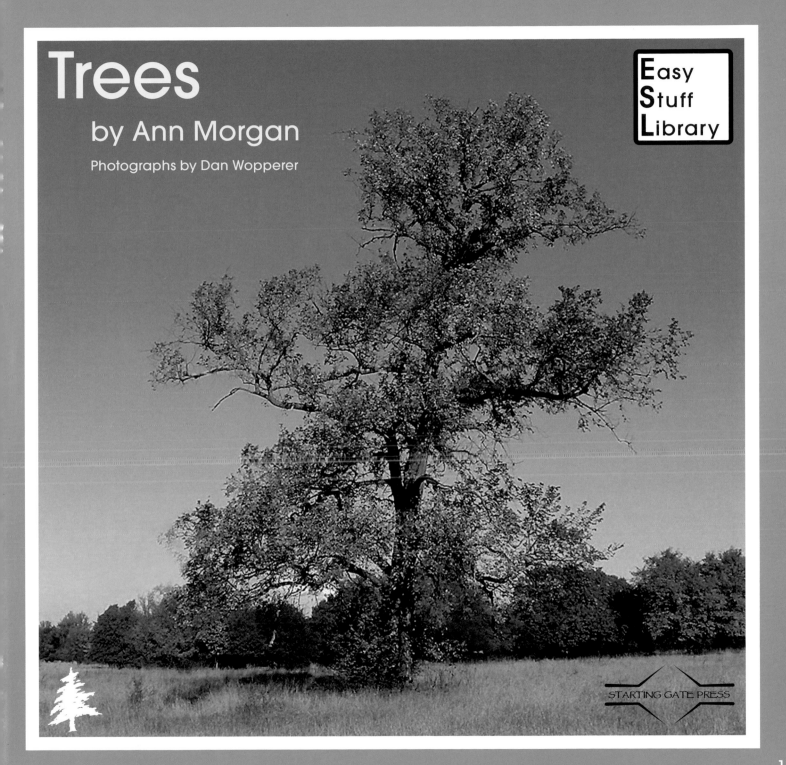

Trees

by Ann Morgan

Photographs by Dan Wopperer

Easy Stuff Library

STARTING GATE PRESS

This series is dedicated to the loving memories of our mothers,
Maureen Cotter Morgan and Dorothy Coleman Wopperer

The author and illustrator would like to extend their heartfelt thanks to the following people for their enduring enthusiasm and encouragement for this project, as well as for their professional expertise:

Nicole Wopperer and Eric Willis, IT design and layout, RareEdge Design Group
Patrice Morgan, MBA
Carl R. Sams II and Jean Stoick
Janet Lasky
Maureen Gardner
The Morgan Girls – Putt, Peggy, Susan, Patrice
The Wopperer Girls – Christy, Nicole, Jan, Erin
Ellen Morgan Rose and Peter

First Edition
Library of Congress Cataloging-in-Publication Date
10 9 8 7 6 5 4 3 2

Morgan, Ann. Trees/ by Ann Morgan, photographs by Dan Wopperer

Summary: The natural beauty and unique features of trees are presented in a simple text format for emerging readers and speakers of the English language, with supporting vivid photographic illustrations.

ISBN 0-9773253-090000
Easy Stuff Library (ESL) Series
Starting Gate Press - startinggatepress.com
Printed and bound in Canada by Friesens of Altona, Manitoba

Table of Contents

Word Help – Before You Read

Words to Practice

roots	veins	ripe
shoot system	needle	fruit
leaves	coniferous	nuts
trunk	broad leaf	cones
branches	deciduous	wings
stem	flowers	husk
limbs	pollen	shell
bark	seeds	scales

Word Help

More Words to Practice

Plural Words

one branch	two or more branches
one leaf	two or more leaves

Compound Words

inside	=	in + side
evergreen	=	ever + green
broadleaf	=	broad + leaf
outside	=	out + side
needleleaf	=	needle + leaf

Introduction

Trees are all around us. Trees can grow in almost every place in the world. There are many different kinds of trees. Every tree has three parts. Every tree has:

roots

a shoot system

leaves

Roots

Every tree has roots. Roots grow under the tree into the ground. They grow all the time. Roots help the tree get food so it can grow. Roots keep the tree in one place. Roots drink water for the tree. Every tree has to have water. Every tree has roots.

Shoot System

Every tree has a shoot system. The shoot system is the trunk and the branches. The trunk is the big stem made of wood. The trunk makes the tree stand up. Small parts from the trunk stick out to the sides. They are the branches. Branches are limbs.

The trunk and branches have bark. Bark is like skin for a tree. Bark covers all of the trunk and branches. Bark keeps the inside of the tree safe so it can grow. Every tree has a shoot system.

Leaves

Every tree has leaves. Leaves are flat and grow out of the branches. A thin stem holds each leaf to a branch. Each branch can have many, many leaves. Most of the time the leaves are green. Every leaf has veins. Veins are like small straws inside the leaf. Water moves up the trunk out into the veins.

There are two kinds of leaves. They are needle leaves and broad leaves. A needle is a very thin leaf. Needles stay on the tree most of the time. Trees that have needles are evergreen trees. Needles almost always look green. Evergreen trees grow new green needles before they drop old brown ones. Evergreen trees are coniferous trees.

A wide leaf is a broad leaf. Broad leaves change color when it starts to get cold and dark. They turn from green to yellow, orange, red and brown. Broad leaves fall off the tree before winter. New green broad leaves grow on the tree after winter. Broadleaf trees are deciduous trees. Needle leaves and broad leaves grow in many sizes and shapes. Every tree has leaves.

Flowers

Every broadleaf tree has flowers. Broadleaf flowers come in many sizes. Some broadleaf flowers are very small and hard to see. Some broadleaf flowers are very big and easy to see. Every flower has pollen to make seeds. Every broadleaf tree has flowers.

Seeds

Every tree has seeds. Seeds grow on the tree until they are ripe. Ripe seeds fall to the ground when they are ready to grow. Ripe seeds can grow into new trees. Seeds have covers around them. The covers for the seeds can be fruit, nuts or cones.

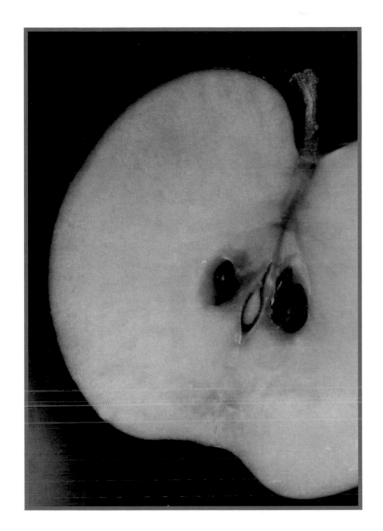

Some seeds grow inside fruit. Seeds in fruit are easy to see when the fruit is open. Fruit is easy to open. Some kinds of fruit have only one big seed inside. Some kinds of fruit have many small seeds inside. Some kinds of seeds have little wings on the outside. A wing helps each seed fly to the ground when it is ready to grow.

Some seeds grow inside nuts. Each nut has a very hard cover. The cover is a husk or shell. The cover goes all the way around the seed. Nuts are hard to open. Each nut has only one seed inside.

Fruit and nuts grow on broadleaf trees. Fruit and nuts do not grow on needleleaf trees. Needleleaf trees grow seeds in cones. Cones have many sizes and shapes. Cones hide seeds under very small covers. The covers are scales. Scales are all over each cone. Each scale hides a seed like a blanket. Ripe seeds fall out of the scales. All trees have seeds that grow in fruit, nuts or cones. Some seeds grow into new trees. Many seeds are food for animals.

29

Summary

Every tree grows from a seed. Some trees have needle leaves and some trees have broad leaves. Some trees have flowers and some trees do not have flowers. Trees can grow seeds in fruit, nuts or cones. Trees can grow in almost any place in the world. Trees are all around us.

Trees Chart

	Broadleaf Trees Deciduous Trees	Needleleaf Trees Coniferous Trees
roots	yes	yes
shoot system		
trunk	yes	yes
branches	yes	yes
bark	yes	yes
leaves	yes	yes
flowers	yes	no
pollen	yes	yes
seeds	yes	yes
fruits	yes	no
nuts	yes	no
cones	no	yes

More Pictures

Some trees grow on tree farms. Tree farms grow trees for people to buy.

Palms have shoot systems different from deciduous and coniferous trees.

People can use trees to make beautiful gardens and parks.

Coniferous trees keep their needles in winter.

Some trees make juice in early spring. The juice is sap.

Deciduous trees get new leaves and flowers in spring.

New leaves start from branches. New leaves are shoots.

Most deciduous trees have green leaves in summer.

Some deciduous trees do not have green leaves in summer.

The leaves on deciduous trees change color in fall.

The leaves on a tree can change color at different times.

Deciduous trees drop their leaves in fall.

Some trees are very, very small. Some trees are very, very big.

Trees need clean air and room to grow.

People take care of trees in greenhouses. Trees can grow inside a greenhouse.

People can make small places up in trees. The places are treehouses.

Book Talk – After You Read

1. Tell one place you can see trees.
2. Tell two things about seeds.
3. What colors can leaves be?
4. What does a tree need to grow?
5. Tell what a needle leaf looks like.
6. How is tree bark like skin?
7. What does "evergreen" mean?
8. Look at the "More Pictures" pages. Find pictures of three broadleaf trees. Find pictures of three needleleaf trees.
9. Draw a picture of a needleleaf tree or a broadleaf tree. Label three or more parts of your picture.
10. What word about evergreen trees is like the word "coniferous"?
11. Draw a new cover for this book. Make a new title on the cover. Put your name on the cover.
12. Do you like broadleaf or needleleaf trees more? Tell why you like that kind of tree.